Indian Martian Odyssey

A Journey to the Red Planet

Srinivas Laxman

PARTRIDGE

A Penguin Random House Company

To order additional copies of this book, contact
Partridge India
000 800 10062 62
orders.india@partridgepublishing.com

www.partridgepublishing.com/india

Contents

Dedicated to

My Mom and Dad,

My Wife, Usha and of course my

Dearest and Only daughter, Rimanika,

Who has always been a source of Inspiration to me.

Preface

It is perhaps not quite common for someone to do a second book on the same topic within a span of one-and-a-half years after the release of the first book.

My first book, ``Mars Beckons India: The Story Of India's Mission To Mars,'' was released by none other than former Indian President, APJ Abdul Kalam, himself a rocket scientist, during the centenary of the Indian Science Congress held at Kolkata in January 2013. As I was unable to have access to the dais that day because of the huge crowd of school children, I had to stand on the side and clap with the audience when the book was formally launched!

The first book basically deals with the pre-launch aspects of the mission. After this, I covered various aspects of the project and filed stories for the Times of India, India's leading national daily. I continue to do reports for the paper even though I retired from it in October 2009. Therefore, in a way my umbilical cord with this newspaper still remains!

I covered the launch on November 5, 2013, at Sriharikota along with three of my Times colleagues—each of us handled different aspects of it— and I did the actual launch. It was a dream turning into a reality for me.

Subsequent events of the mission clearly suggested yes there was enough material for a second book. After all this flight has evoked global interest and curiosity. Therefore, when the opportunity came from Partridge I did not waste time to grab it at once.

This book essentially deals with the post-launch aspects like the orbit-raising manoeuvres and the trans-Martian insertion. Still, brief references and quotes from the first book seemed inevitable as they both deal with the same subject.

I have cited all the sources like ISRO and NASA and private space groups.

I am a space geek and passionate about Mars explorations and in particular the Indian mission. I have collected several pictures and facts and figures about it which I have got laminated and displayed in my room literally turning it into a mini Mars mission museum!

This apart, I have also got t-shirts and mugs specially made for me with the image of the spacecraft and the words ``Indian Mars Orbiter Mission," printed below it.

In the end I would like to express my gratitude to ISRO chairman, K. Radhakrishnan, Mars mission programme director, Mylswamy Annadurai and its project director, Subbiah Arunan, for sparing time from their tough daily schedule to answer my questions; Isro's director for publications and public relations, Deviprasad Karnik, for being of assistance at all hours of the day, B.R. Guruprasad his colleague for his help and to my friend Jal Taraporevala of the Times of India for useful suggestions and guidance. Thanks also to Kinjal Shah and Tejas Taori for their help.

My sincere thanks also to Gian Reyes and Gemma Ramos of Partridge who have gone out of their way to help me.

Finally this preface would not be complete if I do not acknowledge the role of my wife, Usha, and my daughter, Rimanika, who provided good tips and above all patiently listened when I loudly played the Mars mission launch countdown frequently on my laptop! It is music to my ear.

Srinivas Laxman

March 2014

Mumbai, India

Chapter 1

The Mother of All Slingshots

The night of November 30, 2013. While many in India were in the grip of the Saturday night fever, it was, indeed, a different kind of weekend for the scientists and engineers at the Mars mission operations control room at ISRO's (Indian Space Research Organisation) telemetry, tracking and command network in Peenya, close to Bangalore.

In this hi-tech control room filled with huge display screens and digital clocks, there was a mood of nervous apprehension mixed with a sense of excitement.

A team of scientists were glued to their computers monitoring data from India's first Mars-bound spacecraft designated as the Mars Orbiter Mission or affectionately just known as MOM. It was launched on November 5, 2013 at 2.38 p.m. from the Satish Dhawan Space Centre, Sriharikota, a vast spaceport, near Chennai.

The main role of the maiden mission is to explore the surface features of Mars, its morphology, topography, minerology and the Martian atmosphere with five indigenous instruments.

The scientists were keeping their fingers crossed because a crucial event, which was about to take place, had to go off smoothly. If successful, it would mark an important milestone in India's nearly 50-year-old space history and they were anxiously waiting for this to happen literally sitting at the edge of their chairs that night.

As the important development was nearing the spacecraft at one point was oriented in such a way so as to enable its 440 N liquid apogee motor to start firing or operating.

It was 12.49 a.m. a time when the night's excitement in many parts of India, particularly among the urban youth reached fever pitch at discos, night clubs and restaurants. In the control room the countdown clock with its red digital numbers flashed T-O:00:00 indicating that the much-awaited moment had finally arrived. At that point the motor began its 22-minute long firing imparting an incremental velocity to the spacecraft of 648 metres per second consuming 198 kg of fuel of the total of 852 kg.

This precise velocity is important to fly MOM from the earth's orbit, then operate through the earth's sphere of influence, fly through the influence of the sun also known as the heliocentric phase and then finally come under the grip of the Red Planet--the Mars' sphere of influence which is six lakhs kms away from the surface of the Red Planet in September.

Watching the computers, some of the scientists were praying that this significant exercise that Saturday night and early Sunday should be glitch-free, and yes their prayers were answered. Then at about 1.15 a.m. ---it was now the beginning of Sunday---as soon as the motor's operation had ended someone in the team announced with excitement that the firing was successful and instantly the atmosphere in the control room became somewhat relaxed, and tension and anxiety gave away to a sense of relief and joy. It triggered applause among the scientists and they exchanged handshakes.

THE SCENE OF ACTION: THE MARS MISSION OPERATIONS CONTROL ROOM AT ISRO'S TELEMETRY, TRACKING AND COMMAND NETWORK IN PEENYA, BANGALORE ON THE NIGHT OF THE MOTHER OF ALL SLINGSHOTS: CREDIT ISRO.

But all was not over as yet because a very cautious supervisor attached to a group which determines the precise orbit of the spacecraft did not want to take any chances; he asked his team to recheck MOM's orbit which they did.

It was December 1, 2013 and the time -1.30 a.m. A recheck of the orbit showed that MOM was on the right track. When this was confirmed ISRO officials made a historic announcement that India's prestigious Rs 450-crore mission to Mars was on the correct energy-saving trajectory.

The event dubbed as the ``mother of all slingshots'' is known as the trans-Martian insertion (tmi) which marked the beginning of MOM's 680 million km flight towards the Red Planet lasting for about 300 days by moving away from earth's gravity. It was also the start of India's first major interplanetary voyage.

This historic journey was achieved by a complex combination of navigation and propulsion technologies governed by the gravity of the sun and Mars assisted by the 440N liquid apogee motor.

To celebrate the success and conforming to ISRO tradition laddoos—an Indian sweet-- were distributed to the scientists and engineers in the Mars mission operations control room.

Though the trans-Martian insertion went off smoothly, ISRO scientists did, however, experience some brief moments of anxiety. Seconds before the firing of the liquid engine a thunderstorm struck the Hartebeesthoek ground station in South Africa which had been selected for monitoring the trans-Martian insertion operation.

This had resulted in a data loss causing a five-minute delay in the confirmation of the firing of the engine.

The trans-Martian event is important because with it the earth orbiting phase of MOM had come to an end. It was now on its way to Mars which it is expected to reach in September 2014 when the nail-biting Mars Orbit Insertion exercise will take place--a moment of tremendous challenge.

Sure the very thought of this important manoeuvre is causing anxiety to the Mars team because of the 51 global missions to Mars, only 21 have succeeded.

The trans-Martian insertion was significant in more than one way--- with this event first part of operation of the 440 liquid apogee motor had concluded. This motor has to restart on its own in September for entering the Martian orbit.

This weekend operation the--``mother of all slingshots,''-- was ISRO's first experience in sending a spacecraft beyond the earth's sphere of influence which opened a new chapter of interplanetary missions for this country.

Once the exercise was completed MOM's official face book announced the success. Within minutes it was flooded with congratulatory messages from all over the world. ``We have taken the world with us in this mission,'' ISRO chairman K. Radhakrishnan stated.

If in September MOM makes it to Mars at the very first shot itself, it will mark a new beginning in the history of world space exploration enhancing India's status as a growing global space power.

Chapter 2

The Launch

India's celestial pathway begins at the Satish Dhawan Space Centre, Sriharikota, --about a three-hour drive from Chennai.

Driving northward from Chennai, on the main highway towards Kolkata, the capital of the state of West Bengal, one passes several villages and a few towns. Moving along one comes across any number of randomly-located tea stalls making a fast buck by also selling newspapers and magazines, mostly in the Tamil and Telugu languages of Southern India.

People just hang around them, exchanging gossip, flipping through the pages of the publications or staring at the garish film posters and enjoying the loud film music which emanates from these stalls. The scenario is in stark contrast to the hi-tech area which is not far away.

Progressing further north you cross into the state of Andhra Pradesh and then reach a town called Sullurpetta. Turn right here, get off the highway and move on to a narrow road leading towards the space centre which is a nearly 30-minute drive.

Being the last township before the space complex and constantly under its shadow, a sizeable percentage of the local population at Sullurpetta, a not-too-clean place, is fully acquainted about events in the space sector, not only of those relating to India, but similar developments abroad as well. Have a chat with soft drink, or coconut water or ice cream vendor and you will be surprised by his or her awareness about space programmes.

THE GRAND LIFT OFF: THE LAUNCH OF MARS ORBITER MISSION ON NOVEMBER 5, 2013 AT SRIHARIKOTA: CREDIT ISRO.

It is a picturesque journey from Sullurpetta to Sriharikota passing the famous Pulicat Lake, and nearby is a bird sanctuary. If you are lucky you can see some lovely birds flying around.

As you keep going, gradually at a distance the sillouhettes of different types of structures and the rocket start coming into view. They become larger and larger and then a while later a garden with rocket models greets you. You have reached the Satish Dhawan Space Centre, Sriharikota which is an island.

Situated between the Pulicat Lake and the Bay of Bengal, the space complex covering an area of nearly 43,360 acres with a coastal length of 50 kms became operational on October 9, 1969.

Prior to it becoming a space centre, it was the home for a tribal population known as the Yannadis. When the government decided to acquire the area for making a spaceport, the Yannadis were relocated and rehabilitated without any problem. The place is filled with Eucalyptus and Cassuarina trees and is basically a shrub jungle.

Sriharikota on the eastern coast of India proved to be an ideal choice for constructing the spaceport because the rocket can use the spin of the earth for an added boost and follow the orbital motion of the planet. This in turn conserves rocket fuel. Even NASA's Kennedy Space Centre is located on the eastern coast of the US.

Presently, Sriharikota consists of two launch pads, a newly-built mission control centre, a well equipped media centre, a solid propellant booster plant, a static test and evaluation complex, computers and data processing units, a real time tracking system and a meteorological observation unit just to name a few.

PANORAMIC VIEW OF THE FIRST LAUNCH PAD WITH THE ROCKET DURING LAUNCH REHEARSAL FOR THE MARS MISSION: CREDIT ISRO.

Till January 5, 2014, there have been a total of 41 launches of which 29 have been successful. The ones which succeeded catapulting India into the world space league include the unmanned mission to the moon, Chandrayaan-1, launched on October 22, 2008, and the Mars Orbiter Mission which took off on November 5, 2013.

The weather in the spaceport is a matter of uncertainty in October and November when the north-east monsoon sets in over the region. It is a cyclone-prone area and any mission scheduled for these two months is always accompanied by an element of risk. For example during the Chandrayaan-1 launch in October 2008 to the moon, there was lightening and rain even moments before the launch raising doubts whether the mission would take off on that day.

But, just a few minutes before the lift off, the weather decided not to play spoilt sport, it cleared, the moon came out and the launch took place much to the relief of everyone.

The reason why ISRO undertook an enormous risk in launching its first-ever mission to Mars during this uncertain period weather-wise is purely because of orbital considerations. Prior to lift off, the launch authorisation board gave the green signal for the mission on November 1 and the final countdown started at sharp 6.08 a.m. on November 3.

According to space experts at Cornell University about every 26 months, the earth and Mars reach a position in their respective orbits that provides the best trajectory between the two planets. This best opportunity came in November 2013. Originally, it was to have been October, but was rescheduled to November taking all these factors into consideration.

NASA's mission to Mars called Maven (Mars Atmosphere And Volatile Evolution Mission) which incidentally is also an orbiting mission was launched on November 18, 2013, a fortnight after India began its journey towards the Red Planet.

According to the experts, even timing is important because earth is a moving launch pad and Mars is a moving target. Even though the two planets are travelling in the same direction, they are however moving in different orbits.

Keeping this in view they said that rocket engineers must plot a separate orbit for the spacecraft which connects the position of the earth at launch with that of Mars months later.

The Cornell experts point out that the launch has to be timed in such a way that Mars and the spacecraft converge at exactly the same point in space.

Now back to the Indian Mars mission, heading towards Sriharikota for the launch on November 5, 2013, there were a number of security checks on the road from Sullurpetta, but it was a hassle-free clearance for media persons at the main gate of the space centre.

This mission had attracted worldwide interest, and what better proof of this than the fact that the media centre was overflowing with correspondents both from the print and electronic media, not only from India but from foreign countries as well? There was excitement in the air and they started arriving right from the early hours of the day even though the launch was still a few hours away.

The area was filled with TV vans, and correspondents attached to the electronic media were vying with each other to provide the best coverage of the historic mission.

TV monitors positioned in different locations at the media centre were constantly providing a live account of the pre-launch events at the mission operations control room.

The atmosphere in that room was surprisingly calm and relaxed even though a launch of utmost significance and importance, both nationally and internationally, was about to take place. ISRO chairman, K. Radhakrishnan, was greeting and exchanging handshakes with his colleagues and other visitors who had been invited for the launch---the latter were seated in a special enclosure just behind the control centre.

Fortunately the weather was a 'go' for launch. Twenty four hours prior to the take off a weather bulletin said: "No severe weather expected. During launch partly cloudy weather and no rain."

As the moments ticked away, team members of the Mars mission wearing white coats took their position analysing the continuous flow of information which the computers poured out from the rocket and the Mars orbiter spacecraft. They were seen exchanging data and comparing notes.

The countdown was progressing smoothly and was being relayed over the public address system. This is the way it went.

``All stations please standby for the time mark: mark minus 25 minutes and counting.'' Twenty-five minutes left for the lift off and everyone in the mission control room was asked to deactivate their mobile phones. Five minutes later all ground stations were requested to switch to channel two for communication which was followed by an announcement saying that the mission was a `go' for launch. ``Stage parameters are normal and vehicle is ready,'' came another declaration.

``All stations please standby for the time mark: mark minus 17 minutes and counting: All ground stations are ready to support the PSLV-C25/ Mars Orbiter Mission.'' The range safety officer then stated that the range too was ready for the launch.

``All stations please standby for the time mark: mark minus 15 minutes and counting:'' Mission director P.Kunhi Krishnan declared: ``Based on all the parameters, the launch operations sequence has been authorised for the PSLV-C25/ Mars Orbiter Mission.'' He then activated the launch key.

The vehicle director, B.Jaykumar, then gave the green signal for initiating the automatic launch sequence system. From now onwards the computers assumed control of the launch procedure, and by chance if there was any anomaly they would automatically halt the lift off.

``All stations please standby for the time mark: Mark minus six minutes and counting:" The spacecraft's (Mars orbiter) on board computer flight programme was started. With now just 30 seconds left for the take off the rocket's real time programme was activated.

It was 2.38 pm (ist) Tuesday November 5, 2013. The countdown hit the zero mark. The rocket lifted off carrying the 1350 kg Indian Mars orbiter. Significantly, it was the silver jubilee flight of the rocket—the highly proven four-stage Polar Satellite Launch Vehicle (PSLV).

For the first few seconds, the brown-and-white vehicle could not been seen departing as the view of the launch pad was blocked by the jungle. But then everyone felt a sense of relief when a voice over the loudspeaker said: ``Lift off normal," setting off a huge round of applause among the scientists in the control room.

Initially there was no sound even as the rocket carrying the spacecraft came into sight. But minutes later its steady climb was accompanied by fire and sound. Its awesome ear-deafening roar echoed throughout the area as the rocket shot upwards gaining velocity every second followed by a yellow and white fiery plume, and piercing the clear blue sky over the spaceport. This much-awaited moment was a grand spectacle and perhaps one for the history books.

Most people, who had gathered on the terraces and balconies of buildings in the spaceport clapped, shook hands with each other and some even shed a drop of tear or two having been overcome with emotion.

The scene repeated itself outside the gate of the space centre where a large number of people, including women and children, who had no access to the space complex had gathered. All felt a sense of participation with this mission which had grabbed world headlines.

Standing on the terrace of the media centre, media persons watched the fantastic lift off. Then slowly the rocket began to fade out of sight, becoming a dot in the sky, disappearing into the clouds and finally it could not be seen. And with it, its sound also disappeared.

After about three minutes the heat shield of the rocket separated which was an important moment of the mission.

As it flew further and further it passed through a phase when there was a long 28-minute coasting period over the South Pacific Ocean which was a first for any ISRO mission. There was a critical pre-planned manoeuvre when the rocket switched off its engine on its own to save fuel. That is not all. Of these nerve-wracking 28 minutes it not visible to any ground station for 10 minutes. It was completely a blind phase of the flight making the scientists somewhat nervous.

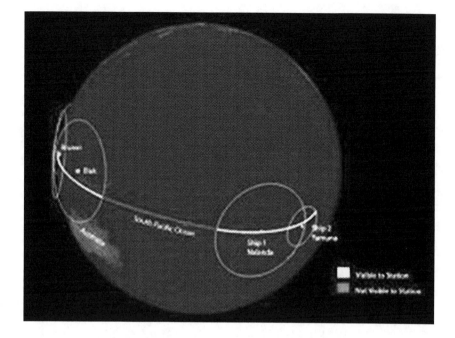

ONE OF THE NERVE-WRACKING MOMENTS: THE FLIGHT
OF THE SPACECRAFT OVER THE SOUTH PACIFIC OCEAN
AFTER LAUNCH ON NOVEMBER 5: CREDIT ISRO.

The long coasting called for specific modification and validation of the steps relating to coast phase guidance, on board battery capacity augmentation and assessment on the performance of the inertial systems for extended flight duration.

This phase of the flight will be also be remembered for another important manoeuvre---the orbiter took an unusual curve of about 275 degrees to achieve what scientists call the ``argument of perigee.'' In previous missions it was only 178 degrees.

This unusual curve was necessary to minimise the energy needed while the spacecraft moves from earth to Mars.

Since there were no ground stations around this area of the South Pacific Ocean ISRO had hired two ships of the Shipping Corporation of India, ``SCI Nalanda,'' and ``SCI Yamuna'' which were equipped with radar and tracking equipment. Again this was the first time that ISRO was depending on civilian ships for supporting a space mission and they operated off the Fiji Island.

The government of Fiji provided all help to these two ships and its crew comprising ISRO scientists as well. As a gesture the scientists presented a model of the Mars orbiter to the Fiji president, RatuEpaliNailatikau.

The two ships closely monitored the mission during the third and fourth stage operations of the rocket, and most importantly the separation of the spacecraft from the rocket 42 minutes after launch.

Therefore, imagine the sense of relief the scientists felt when a while later, it was declared that ``SCI Nalanda'' had confirmed that it had received the signal from the spacecraft'' and all was going well. The message was instantly flashed to the ISRO' telemetry, tracking and command network at Bangalore.

Five minutes after this announcement, and about 2100 seconds into the flight the engine in the spacecraft reactivated itself on its own. And then came the historic moment from Dr Radhakrishnan who declared that the

Mars orbiter had been successfully placed into its initial elliptical earth parking orbit at an altitude of 23,550 kms. He announced that the first part of the mission was a success.

This was nearly 2660 seconds after the launch corresponding to 42 minutes.

The flight was being tracked by the Indian Deep Space Network at Byalalu off the Bangalore-Mysore highway through its 32-metre antenna, and also NASA's Deep Space Network at Goldstone California, Madrid and Canberra in Australia. The 32-metre antenna at Byalalu was upgraded for the Mars mission. Even during the US shutdown NASA's deep space network supported the mission.

THE 32-METRE ANTENNA AT THE INDIAN DEEP SPACE NETWORK AT BYALALU NEAR BANGALORE WHICH IS TRACKING THE MARS ORBITER MISSION: CREDIT ISRO

NASA'S DEEP SPACE NETWORK AT GOLDSTONE IN CALIFORNIA
WHICH IS PROVIDING SUPPORT TO THE MARS ORBITER
MISSION ALONG WITH THE ONES AT MADRID AND
CANBERRA. CREDIT JET PROPULSION LABORATORY NASA

The telemetry data from the launch vehicle was flashed to the mission computer systems at Sriharikota where it was processed

An upcoming US-based sci-fi writer, Andy Weir, who has just come out with a book called "The Martians," which is making waves stated in his face book just after the launch: "another good day for Mars Exploration. India has successfully launched a Mars probe! It's called Mangalyaan and it will sit in Mars orbit, scanning the atmosphere and surface for anything interesting. They have a specific interest in finding methane. Methane can be produced by geological events and by biological life forms. Also it decomposes quickly. Mars is geologically dead, so if we found large amounts of methane in its atmosphere, it would be a heavy indicator of biological processes going on. NASA's Curiosity probe didn't have any luck looking for methane, but it's on the ground. Mangalyaan will be able to do comprehensive scans of the atmosphere from space. India is developing a niche for themselves in the world of orbital chemical analysis. It was their probe Chandrayaan-1 that brought back the first reported data of water on the moon. Shubhyatra Mangalyaan."

Chapter 3

India heads towards the Red Planet

While India went into a celebratory mood reminding one of a big festival after the successful launch, it, however, raised an important question: Why did the Mars orbiter not fly directly to the Red Planet?

Why did the spacecraft have to be initially placed in a parking orbit of 23,550kms and its altitude raised subsequently in phases before it escaped from the earth's gravity, and begin its 300-day flight towards Mars in the early hours of December 1, 2013.

The reason offered by ISRO officials is that though the PSLV is a super rocket and has proved a true and dependable work horse, it has however weight limitations and cannot impart the necessary velocity for the spacecraft to zoom directly towards the Red Planet.

THE SPACECRAFT BEING ENSCAPULATED IN THE
ROCKET'S HEAT SHIELD: CREDIT ISRO.

It is for this reason that after the Mars orbiter was placed in a parking orbit first after the launch, it thereafter executed six orbit raising manoeuvres which increased its agogee (the farthest point from the earth).In this way the spacecraft steadily gained velocity prior to the critical trans-Martian insertion on December 1, 2013.

The mission was so designed based on the low payload capability of the PSLV. In an earlier interview, Dr V. Adimurthy of ISRO who worked out the trajectory explained that the minimal-energy orbit placement around Mars and the optimal utilisation of the existing launch vehicle systems are the main driving factors for the Mars Orbiter Mission Design

THE SPACECRAFT IS ENCLOSED IN THE ROCKET WITH
THE HEATSHIELD CLOSED. CREDIT ISRO.

David Doody, a senior engineer of JPL has drawn a comparison between MOM and Maven since both were launched in the same month.

Quoted by Emily Lakdawalla in her Planetary Society blog, David explains that for Maven the trip to Mars is straightforward. That's because the Atlas 5 rocket was able to deliver Maven which was fully fuelled and an attached powerful Centaur upper stage into earth orbit. Centaur was able to kick Maven directly on to Mars.

But, in contrast ISRO's PSLV could deliver MOM to earth orbit, but not with a powerful upper stage like the Centaur. According to Lakdawalla, it took about five minutes of thrust of the Centaur to enable Maven to depart from the earth orbit for its interplanetary voyage.

For MOM on the other hand she calculated about 40 minutes of total thrust comprising six manoeuvres.

In the first such exercise 72 hours after the launch on November 7, a team of 150 scientists at ISRO's telemetry, tracking and command network in Bangalore began the operation an hour after midnight at 1.17 a.m. The liquid apogee motor fired for 416 seconds which raised the altitude of the spacecraft from 23,550 kms to 28.825 kms consuming 40 kgs of fuel.

When this happened the Mars team wanted to raise a toast because as it reached an altitude of 28,825 kms it broke a world space record too. The mission became the second most tracked one globally just three days after it took off and even got a five-star rating, according to a real time satellite tracking data.

According to this data, the first position went to the European Space Agency's Gravity Field and Steady State Ocean Circulation Explorer Satellite.

Even as these honours were being bestowed on the Mars Orbiter Mission a second orbit raising manoeuvre was carried out on November 8 at 2.18 a.m. by operating the liquid apogee motor for 570.6 seconds which further raised the spacecraft from 28,825 kms to 40,186, kms—an operation which utilised 83 kgs of fuel.

The third exercise on November 9 held at 2.10 a.m. took the spacecraft further up to an altitude of 71,623 kms. For this increase the engine was fired for 707 seconds.

But the scientists experienced some disappointment when there was a setback in the fourth orbit raising drill on November 11.

Original, plans envisaged taking the spacecraft up to a height of one lakh kms. But, this did not happen because the motor did not burn long enough on account of a minor technical glitch. Consequently, the spacecraft could be taken only to an altitude of 78,276 kms. This imparted an incremental velocity of 35 metres per second as against 130 metres.

The situation turned a bit scary because the scientists lost track of the spacecraft for about 15 minutes. To their relief they reacquired it, locked on to it and confirmed that it was after all MOM. Everything was fine thereafter.

During these operations which kicked off on November 7 the space agency had been evaluating the spacecraft's autonomy functions which were essential for the trans-Martian insertion and the Mars capture in September.

In the first three exercises, the spacecraft's prime and redundant chains of gyros, accelerometers, the 22 Newton attitude control thrusters, the attitude and orbit control electronics as well as those connected with fault detection isolation and reconfiguration were tested satisfactorily.

The scientists were happy that both the star sensors---the prime and redundant one—were also functioning normally.

In the fourth orbit raising drill on November 11 different types of redundancies relating to the propulsion system were tested. But, then why did the small setback occur?

Explaining why the spacecraft's altitude could not be raised to one lakh kms, ISRO said when both the primary and redundant coils of the engine were energised simultaneously—an exercise which had been planned earlier---the flow of fuel to the liquid engine stopped.

According to ISRO though the operation continued using attitude control thrusters, it was not without a minor penalty—the sequence caused a drop in the orbiter's incremental velocity. But, it had no impact on the mission as a whole. Thereafter, the flight to Mars continued glitch-free.

Three days later after the problem was rectified, the orbiter successfully zoomed into the deep space zone on November 12 crossing an important milestone. The exercise began around 5 a.m. and the engine fired for 303.8 seconds. This was the fifth orbit raising manoeuvre taking the spacecraft's altitude from 78,276 kms to 1, 18,642kms.Isro,however prefers to describe it as a fourth supplementary manoeuvre.

About the precise point where the deep space zone starts, the scientists maintained that any point in space a little after one lakh km marks its beginning.

November 16 was important for the Indian Mars mission because on this day the orbiter took a giant leap. Its altitude shot up by nearly a lakh kms which was the biggest since its launch on November 5.

This was the sixth and final drill to raise the altitude of the orbiter and the command for firing the engine was flashed from the control room at 1.27 a.m. and it operated for 243.5 seconds. With this the spacecraft went up from 1, 18,642kms to 1, 92, 874 kms. This was the last such exercise before the trans-Martian insertion on December 1.

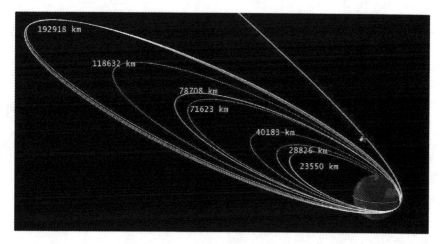

THE ORBIT RAISING MANOUVRE OF THE MARS ORBITER MISSION
BEFORE THE TRANS MARTIAN INSERTION: CREDIT ISRO.

After successfully executing all these manoeuvres, the orbiter became the farthest Indian object in space when it crossed the lunar orbit on December 2 which is approximately 3,85,000 kms away from earth. And then it bade its final goodbye to earth an hour after midnight on December 4when it flew beyond the earth's sphere of influence which extends to about 9,25,000 kms.

Exactly a week later the spacecraft executed what is known as a trajectory correction manoeuvre on December 11—the first of the four---to eliminate any possible deviations in the flight path when the spacecraft was 2.9 million kms away from earth. The 22 Newton thrusters of the spacecraft were fired for 40.5 seconds.

Since the distance between the spacecraft and the ground station was vast, the round signal transmission time was about 20 seconds. The entire operation was thus managed by the on-board computer. Three more such manoeuvres have been planned in June, August and September.

About two months after the first trajectory correction manoeuvre, the spacecraft hit a century on February 12 when it had flown 190 million kms of the total of 680 million kms. On this day it completed 100 days of flying. About 300 kg of fuel are available which is enough for the remaining part of the mission is.

Then almost two months later at exactly 9.50 a.m. on Wednesday April 9 when majority in India were in a rush to leave for their places of work, MOM quietly hit another milestone----the spacecraft crossed the half way point in her 300-day flight to the Red Planet.

Midway to the
Red planet

Today (April 09, 2014) at 9:50 am IST, MOM crossed the half-way mark of its journey to the Red Planet along the designated helio-centric trajectory.

THE ANNOUNCEMENT IN ISRO'S OFFICIAL MARS ORBITER MISSION FACEBOOK STATING THAT THE SPACECRAFT HAD CROSSED THE HALF-WAY MARK TO THE RED PLANET: CREDIT ISRO.

At that moment, it had flown 337.5 million kms in its elliptical orbit around the sun. Coincidentally, this happened just a day after the closest approach of Mars and earth in their respective orbits known as an opposition.

When its altitude was being raised, the spacecraft successfully sustained several passes of lethal radiation belts speculated to have been formed by furious solar winds and harmful cosmic rays. But the spacecraft has been designed to withstand such dangers.

The mission has been split into three phases. These are the geocentric phase, the heliocentric phase and the Martian phase.

In the geocentric phase the spacecraft was under the earth's sphere of influence. To ensure that the spacecraft consumed least amount of fuel, ISRO used what is known as the Hohmann Transfer Orbit or a Minimum Energy Transfer Orbit.

After the geocentric phase the spacecraft flew into the heliocentric phase—the second part of the flight--when it came under the sun's influence. This is a long flight path and is almost one half of an ellipse around the sun.

During this phase, the spacecraft was also under the influence of other planets as well. Significantly, the spacecraft does not need any fuel while flying from earth to Mars in the Mars transfer trajectory. A comment in ISRO's highly popular Face book stated: ``Just like a spacecraft orbits earth without spending any fuel, MOM (Mars Orbiter Mission) is presently circling the sun in an elliptical orbit till it encounters Mars. A little amount of fuel however is needed though for correcting its trajectory and controlling its orientation," it said.

The third phase of the mission for the spacecraft is the Mars sphere of influence which is around 5, 73,473kms from the planet's surface. As the spacecraft begins its closest approach to Mars known as Periapsis it is captured into a planned orbit around the Red Planet through the Mars

Orbit Insertion. Provisionally, this is slated to occur on September 24 about 48 hours after NASA's Maven reaches Mars on September 22.

According to Lakdawalla, both MOM and Maven will get just one chance to enter Mars orbit. MOM will depend upon its single 440 Newton liquid apogee motor, while Maven has six rockets each of which can achieve 200 Newtons of thrust.

Interestingly, although the liquid apogee motor has flown on Insat-series of satellites and the Chandrayaan-1, the engine has, however, never had such a long period of inactivity. It stopped operating after the trans-Martian insertion on December 1 and will restart on its own for the Mars Orbit Insertion 300 days later in September 2014. In fact one of the technological objectives of the mission is to accomplish the Mars capture once the engine revives. Remember, European Space Agency's comet-hunting Rosetta woke up after 31 months of deep space slumber on January 20, 2014.

Mars is more or less 200 times the distance of the moon which means that the challenges and risks for a mission to the Red Planet are greater compared to a flight to the moon.

The orbital mechanics of a Mars mission will be influenced by the Earth, Mars and Sun. Venus is also sufficiently close to exert its influence. So, while the trans-Martian insertion went off by and large smoothly on December 1 one cannot perhaps not rule out the sun and Venus leaving its impact on the flight. But ISRO scientists are confident that MOM will reach Mars without encountering any major problem.

Director of NASA's Jet Propulsion Laboratory, Charles Elaichi, has been quoted as comparing the Mars Orbiter Mission navigation challenge to hitting a golf ball from India to a hole in Los Angeles. According to him, the ball has to come straight into the hole which is moving. This makes the navigation task all the more challenging.

With five scientific instruments MOM is getting closer each day to its hole---which in this case is the Red Planet.

Chapter 4

The Rocket, The Spacecraft and The Payloads.

Who can ever forget the sight of the mighty rocket blasting off at Sriharikota on the afternoon of November 5, 2013 carrying the Indian Mars Orbiter with its five payloads?

The hundreds who were present at Sriharikota that day, and of course the lakhs who were watching the lift on TV all over India and elsewhere too, will always remember this grand spectacle of India zooming towards a new and challenging era not only in its space history, but that of the country as well. It was a moment which has got deeply embedded in one's memory.

The launch of this 44.4 metre tall brown and white rocket, known as the Polar Satellite Launch Vehicle (PSLV), has always instantly set off applause among the scientists, visitors and onlookers because it has proved to be ISRO's reliable work horse fetching a number of foreign customers for the space agency.

This rocket has notched up 24 successful launches since its maiden mission on September 20, 1993---this first flight, however, failed because of an attitude control problem. After this setback it has been a story of one success after another.

The vehicle has so far 40 foreign satellites and 30 indigenous ones.

This highly-successful vehicle was developed in the early 1990s at ISRO's Vikram Sarabhai Space Centre in Thiruvanathanapuram, the capital of Kerala, in Southern India.

With a lift off weight of nearly 295 tonnes, the standard version of this rocket can place 1600 kg class of satellites in the 620 km sun synchronous polar orbit, and 1050 kg class of satellites in the geo synchronous transfer orbit.

The booster stage of the four-stage rocket along with the strap-on motors and the third stage have solid propellants, while the second and fourth stages operate with liquid propellants.

The variant of the rocket which carried the 1345 kg Mars Orbiter was an advanced version of the vehicle designated as the PSLV-XL. This particular type, which had its successful inaugural run, with the Chandrayaan-1 mission on October 22, 2008, has got stretched strap-on boosters to achieve higher payload capability. The enhanced boosters use 12 tonnes of solid propellants as against nine in the normal variants of the rocket. The lift off weight of this version of the rocket is 320 tonnes.

For the Mars mission, it was the 25th flight of the PSLV marking its silver jubilee, and the fifth in the XL configuration. The rocket's mission was designated as PSLV-C25.

Apart from the mission to the moon and Mars, the other four missions which used the PSLV-XL were the launch of a communication satellite, GSat-12 on July 15 2011, flying Risat-1 the heaviest satellite weighing 1858 kg ever to have been placed in orbit by the PSLV on April 26, 2012 and launching India's first and second regional navigation satellites IRNSS-IA and IRNSS-IB on July 1, 2013and April 4, 2014 respectively

ISRO'S WORK HORSE. A FILE PICTURE OF THE POLAR
SATELLITE LAUNCH VEHICLE: CREDIT ISRO.

So much about the rocket and now about its passenger---the spacecraft which significantly was readied even before the government gave its formal go ahead for the mission—a clear indication that ISRO was in a 'go' mode with regards to the flight to the Red Planet.

Manufactured by the aerospace division of Hindustan Aeronautics Limited (HAL) at Bangalore, it was handed over to ISRO on June 25 2012 ---the government approved the project weeks later on August 3, 2012. What better proof that ISRO was all set for the Mars mission?

THE MARS ORBITER MISSION SPACECRAFT: CREDIT ISRO.

A fortnight afterwards on August 15, 2012, PM ManMohan Singh announced this flight during his Independence Day address from the ramparts of the Red Fort in New Delhi. He told the large audience:`` Under the mission, our spaceship will go near Mars and collect important scientific information. The spaceship to Mars will be a huge scientific step for us in the area of science and technology," he stated.

In stark contrast to the way former PM AtalBehari Vajpayee announced the the first Indian lunar mission and President John F.Kennedy declared US's plan to land a man on the moon, there was unfortunately no thrill and excitement in the voice of Singh when he announced the journey to Mars. He mentioned it as though it was just another scientific project!

HAL made the spacecraft in just three months and it cost just 10 per cent of the total project cost. Once readied, it was shifted from HAL to the Isro Satellite Centre in Bangalore under heavy security cover and placed in spacecraft check out room number one which became a star attraction among many ISRO scientists who were vying with each other to have a peep of the Mars orbiter.

Made of aluminium and carbon fibre, it is cuboid in shape and weighs 1,350 kgs. Of this, 850 kgs consists of the propellant and 500 kgs is the dry mass which includes the five payloads totally weighing about 15 kgs.

The propulsion system has two spherical tanks each holding 390 litres of propellant. The spacecraft uses Unsymmetrical Dimethylhydrazine as fuel and mixed oxides of Nitrogen as oxidiser. The main propulsion system is of course the liquid apogee motor.

THE SPACECRAFT'S PROPELLANT TANK BEING READIED: CREDIT ISRO

According to ISRO, the configuration of the orbiter is a balanced mix of design of the flight proven Indian Remote Sensing, the Insat (Indian National Satellite System) and Chandrayaan spacecrafts.

Keeping in view the extremely challenging nature of the Mars flight, modifications had to be made in critical areas such as communication, power, and propulsion and on board autonomy. The spacecraft is equipped with three antennas. The high gain antenna transmits signals up to a range of 400 million kms, the limit of the medium gain antenna is 200 kms and that of the low gain antenna, seven kms.

THE SPACECRAFT'S HIGH GAIN ANTENNA
DEPLOYMENT TEST: CREDIT ISRO.

The spacecraft has eight 22N thrusters which play a major role especially during the trajectory correction manoeuvres.

The orbiter has a single deployable solar array with three panels each measuring 1800 X 1400 mm which compensate for the lower solar irradiance. In the earth-bound orbit, the power generated by them was 1800 watts which is expected to come down to about 900 watts as the orbiter gets close to Mars.

A single 36 amp hour Li-Ion battery is adequate enough to take care of the eclipses in the Martian orbit.

The electrical power provided by the solar array at Mars is channelled to a distribution unit which in turn supplies power to the various systems and payloads and controls the battery for night operations.

The five instruments on the spacecraft, chosen after a rigorous evaluation and a transparent process, will observe the Martian surface, atmosphere and exosphere extending upto 80,000 kms. Data obtained from these payloads is expected to provide a detailed understanding of the evolution of the Red Planet.

The five payloads are the (1) Lyman Alpha Photometer which will study the escape process of the Martian upper atmosphere through deuterium/hydrogen. Weighing 1.97 kg, an analysis of its data will also help scientists the loss process of water from the planet.

The principal investigator of this instrument, L.M. Viswanathan of the Laboratory For Electro Optics Systems in Bangalore, where the instrument was designed, in a joint report with his team members to Planex, a journal of the Ahmedabad-based Physical Research Laboratory has stated: ``it is the first Indian space-based photometer developed utilising the absorption gas-cell technique." Its observation time per orbit will be 150 minutes.

(2) Methane Sensor for Mars. Of all the five instruments information gathered from this payload weighing 2.94 kg will be of considerable significance and interest because it could throw light about the source

of the elusive Martian methane---whether it is biological or geological. If biological it could answer an age-old question—Is there life on Mars?

A few months ago an analysis of the data from NASA's Curiosity rover, which landed on the Red Planet on August 6, 2012, revealed that there was no methane on Mars, stirring up a lot of controversy among space scientists all over the world.

Responding to this, former chairman of ISRO, U.R. Rao, who had played a key role in the selection of the five payloads for the Mars Orbiter Mission dismissed it as simply baseless emphasising that it did not carry much weight.

According to him the value of NASA's announcement was of limited value because Curiosity focussed only on a single area of Mars. The Indian mission on the other hand will look for methane all over the planet since it being an orbiting one.

Writing a collaborative report about the instrument in Planex with his team, its prinicipal investigator, Kurian Mathew, stated that ``MSM (Methane Sensor For Mars) can map the sources and sinks of methane by scanning the full Martian disk from apogee position of Mars Orbiter.''

Developed at ISRO's Space Application Centre in Ahmedabad, he has written saying that the instrument had undergone all kinds of functionality checks during the earth-bound phase of the mission. 'It was found that all payload health parameters are within the specified limits."

(3) The Mars Exospheric Neutral Composition Analyser (MENCA) developed by the Space Physics Laboratory of ISRO's Vikram Sarabhai Space Centre in Thiruvanathanapuram will study the neutral composition of the Martian upper atmosphere. Its weight is 3.56 kg.

Its principal investigator, Anil Bhardwaj, writes in the journal: 'It is believed that the young Mars had an atmosphere substantially thicker than what it is today; sufficient to retain water in its liquid form. The data from the previous and current missions to Mars indicate possible water flow on the surface of the Red Planet. Due to various thermal and non-thermal processes, Mars had lost its atmosphere deserting it in its present form. Study of the exosphere by MENCA may help in understanding the thermal escape of the Martian atmosphere," Bhardwaj has stated.

He said the instrument was successfully commissioned on November 13, 2013, and operated for an hour while the spacecraft was in the earth-bound orbit at a distance of around 70,000 km from the earth. "Its health was found to be normal. MENCA will be operated several times during the 10-month journey of mission to Mars before insertion into Martian orbit on September 24, 2014," he writes.

(4) The Mars Colour Camera developed at the Space Applications Centre will undertake optical imaging. Weighing 1.27 kgs, it will provide information and images about the surface features of Mars and throw light about its composition. Additionally it will be used for studying the two satellites of Mars—Phobos and Deimos.

The camera's principal investigator, A.S. Arya, has written in Planex that it is a medium resolution camera designed to be versatile and multipurpose.

He said that the aspects of the Red Planet which the camera will image will include craters, mountains, valleys, sedimentary features and various volcanic features.

According to him the camera will also be used for studying the Martian polar ice caps and its seasonal variations using its imageries.

Arya has stated that the Martian surface has been imaged during previous missions by other spacecrafts' instruments with better spatial and spectral resolutions. According to him, the highly dynamic nature of the Martian atmosphere and surface requires that every mission has its own imaging payload. "MOM has uniqueness in terms of its highly elliptical orbit," he has stated.

The camera was inaugurated on November 19, 2013, when it imaged Cyclone Helen hurtling towards the coast of Andhra Pradesh. When the camera took a shot of the cyclone, the spacecraft was flying at an altitude of 67,975 kms.

THE FIRST PICTURE TAKEN BY THE MARS COLOUR
CAMERA ON NOVEMBER 19, 2013: CREDIT ISRO.

(5) The Thermal Infrared Spectrometer (TIS) weighing 3.2 kgs, again a product of the Ahmedabad-based Space Applications Centre will map the surface composition and minerology of Mars.

The instrument's principal investigator, R.P. Singh, states in Planex that the payload will also estimate the ground temperature of Mars' surface and detect and study the variability of aerosol/dust in the Martian atmosphere.

He said that data obtained from TIS will be of significance because ``knowledge on the type of minerals present in any planetary system provides information on the conditions under which minerals are formed and process by which they are weathered."

The science data transmitted from the payloads will be stored and archived at ISRO's Indian Space Science Data Centre situated in the campus of the Indian Deep Space Network. This is located at Byalalu off the Bangalore-Mysore highway.

The data will be accessed by the principal investigators of the payloads and later by students and scientists.

The original Indian Mars mission plans envisaged having nine instruments totally weighing 25 kg, if the rocket had been the three stage Geo Synchronous Satellite Launch Vehicle (GSLV) which is capable of carrying more weight.

THE ONCE NAUGHTY BOY! A FILE PICTURE OF THE GEO SYNCHRONOUS SATELLITE LAUNCH VEHICLE: CREDIT ISRO.

However, till January 5, 2014, it had earned a reputation of being a "naughty boy," on account of its somewhat erratic performance. So rightly the Indian space community did not want to risk using this rocket for the country's first Mars shot.

So, the GSLV was replaced by the "good boy," –the highly proven PSLV. But, this "good boy" imposed weight limitations as a result of which four payloads had to be removed. These are expected to be flown on India's second mission to Mars provisionally slated for launch in January 2016, if it materialises.

• (A lot of details have been taken from ISRO documents.)

Chapter 5

The Major Challenges of the Mars Orbiter Mission

Sure, MOM (Mars Orbiter Mission), now well on its way to the Red Planet, will enjoy a tremendous amount of independence and freedom in the days ahead, but it has to behave responsibly!

The spacecraft will approximately cover a total distance of nearly 6, 80,000 million kms by the time it nears the Red Planet. It is a huge distance making it impossible to command it from ground.

Communication with the spacecraft, therefore, is one of the major challenges of this mission. The range between the ground stations and the spacecraft during the hair-splitting Mars orbit insertion on September 24, 2014, will be vast.

Srinivas Laxman

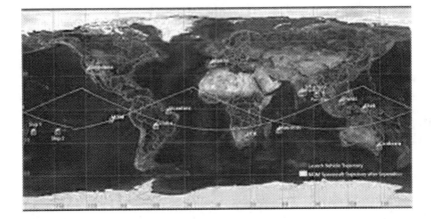

MARS ORBITER MISSION: GLOBAL TRACKING NETWORK: CREDIT ISRO

The first signal indicating the status of the orbit insertion is expected to take approximately 20 minutes to travel from the spacecraft to the ground station, and there is a possibility that the Canberra Deep Space Network in Australia will be the first to acquire this much-awaited signal.

The maximum earth-to-Mars round trip light time is as much as 42 minutes during the mission leading to a lot of autonomy being incorporated into the space craft. What does this indicate? It means it has to take decisions on its own and respond to emergencies without intervention from the ground stations. So a lot of trust and faith has been reposed on the spacecraft that it will not perform erratically!

The spacecraft's on-board autonomy has been implemented through what is known as ``autonomous fault detection, isolation and reconfiguration logics."

According to ISRO, the mission requirements called for the development of 22 software modules which were extensively tested.

THE MARS ORBITER IS BEING PREPARED FOR THE VIBRATION TEST AT ISRO'S SATELLITE CENTRE IN BANGALORE: CREDIT ISRO.

The spacecraft's propulsion system poses another challenge because the liquid apogee motor has to restart on its own after a gap of 300 days for the Mars Orbit Insertion. A lot of safety and redundancy measures have been provided, and the hope is that the insertion exercise will be a smooth affair.

To ensure a flawless performance two liquid engines were tested under simulated conditions at a high altitude test facility at ISRO's Liquid Propulsion Systems Centre in Mahendragiri, Tamilnadu.

Radiation also presents a challenge to the mission. According to Isro here exists the Van Allen radiation belt consisting of two dough net-shaped blankets that cover the earth with highly charged plasma particles consisting of electrons, protons and nuclei.

Prolonged exposure to these belts poses a significant threat to various components of a spacecraft.

Fortunately, however, MOM has survived several passes of these dangerous radiation belts formed by furious solar winds and harmful cosmic rays. These belts are part of the earth's inner magnetosphere and stretch from an altitude of 1000 km to 60,000 km above the earth.

Pointing out that MOM has been designed with sufficient safeguards against such particles, ISRO has stated that the spacecraft's main frame bus elements and payloads have been developed in such a way so that there is generally a trouble-free operation during the earth burns, the Mars transfer trajectory and in the Mars orbit.

This apart, the spacecraft has to cope with a wide range of thermal environment conditions in the near-earth zone because of the sun and earth, to Mars conditions where eclipses and reduced solar flux will give rise to cold case issues.

A research paper about thermal environment states that ``knowledge of planetary environments is needed to predict the thermal loads in planetary orbits and flybys, both for unmanned spacecraft and human exploration.''

Prior to launch, the spacecraft was subjected to what is known as thermal balance tests to simulate Mars conditions, and it came out with flying colours!

One of the major challenges for this mission is the design of the power system in the spacecraft because of its enormous distance from the sun. The power generation in Mars orbit will be reduced between 35 and 50 per cent as compared to the earth orbit. This apart, the solar panels will experience low temperatures during the eclipse period going down to as much as minus 185 degrees C.

The array has been made in such a way that it will ensure maximum performance during the Martian solar flux conditions generating 840 watts of power.

THE SPACECRAFT'S SOLAR PANEL UNDERGOING
PRE LAUNCH TESTS: CREDIT ISRO

A mission to Mars always throws up severe challenges for scientists and engineers. Much of these are known to them in advance while the mission is in its early planning stages, and as a result systems are designed to respond to them.

But, there is always the unknown which suddenly confronts the spacecraft team, very often when the flight is in progress, and it is here they need to exercise their ingenenuity, tackle the problem and ensure a smooth mission.

When NASA's rover ``Spirit'' landed on Mars on January 4, 2004, the team faced a crisis—a few days after its touchdown the spacecraft stopped communicating and there was no was data available. The mission's principal investigator, Steve Squyres, has written in his book ``Roving Mars,'' stating that it was the first major in-flight anomaly. But, to the relief of the team, Spirit awoke and was back in action sometime later. This was the first major challenged they faced.

And here is another famous example---who can forget the nerve-wracking landing of NASA's nearly one tonne Curiosity rover on Mars on August 6, 2012?

Its touchdown was a complex exercise since it involved the use of the sky crane in the final moments of the landing. The use of the sky crane understandably triggered apprehensions that after all with all the effort put in, it may not work at the end and this much-publicised mission may flop.

But, the sky crane operated flawlessly, and the Curiosity mission rocketed into history books. But, this is not to say that in the final phase of the landing there were not minor hiccups which gave some challenge to the Curiosity team. These, were however sorted out and the project proved a super success.

AN ARTIST'S IMAGE OF THE HISTORIC HAIR RAISING TOUCHDOWN OF CURIOSITY ON THE RED PLANET ON AUGUST 6 2012 SHOWING THE SKYCRANE: CREDIT JET PROPULSION LABORATORY NASA.

A Mars mission is known to give sleepless nights and even perhaps nervous breakdown to its team. But, when the story of its success grabs the headlines, they realise after all it is worth all the agony and sacrifice. Mars after all is not too friendly a planet.

Chapter 6

Some of those who made it happen

As MOM is zooming towards the Red Planet, there is alteast one person who has almost become a stranger to his family!

For the last 15 months he has been returning home from his work centre at Bangalore during an unearthly hour of around 3.30 a.m. ---just at the crack of dawn—only to go back a few hours later after attending to his religious duties. His sole focus is on MOM and to ensure that by and large it has a trouble-free flight. That is his dream.

``Very often those at home do not know when I come and go back to work," the scientist remarked in a very light hearted manner.

Who is this person who has been maintaining a punishing work schedule? He is Subbiah Arunan, the mission's project director who has proved an excellent team leader.

Ever since the mission got a 'go' in August 2012 the highly soft spoken Arunan has literally made the ISRO Satellite Centre in Bangalore his home.

He has been sleeping there, and when time and work permit, which is of course very rare, enjoys a P.G. Wodehouse novel which he says helps to de-stress him!

``I decided to stay at the centre to respond to any possible emergencies, attend to mission calibrations and tackle any unforeseen issues, '' he told this writer in a recent interview.

This very amiable team leader said that after the launch the most scary moment for him was during the trans-Martian insertion. ``It was a heart breaking moment and I admit I was very nervous because earlier missions to Mars of other countries had failed during this crucial phase.''

``I was extremely relieved that it went off smoothly. By chance, if this had not succeeded then our spacecraft would have become just another celestial object flying in the solar system,'' he said.

According to him, the Mars Orbit Insertion which will take place on September 24, 2014, at 7.30 a.m. will be less scary than the trans-Martian insertion. ``The spacecraft will enter the Mars sphere of influence on September 21. This means it will be leaving the grip of the Sun that day. As of now I am confident that the insertion, during which the liquid apogee motor will fire for 1400 seconds, will be a perfect exercise and go off smoothly,'' he stated.

He said that the spacecraft currently was flying at velocity of 29.1 kms per second with respect to the sun and on average covering a distance of nearly 1.5 million kms each day.

Once in Mars orbit, the apogee of the spacecraft—the furthest point it will operate--will be 80,000 kms, and the nearest is 366 kms known as perigee. It will be an elliptical orbit.

Arunan has been quoted in a magazine (Hindustan Brunch) interview saying that he was living his dreams. He acknowledged that he was slightly sceptical when the team was given such a short notice to realise the mission.

``But, we conceived, designed, fabricated tested and launched it, everything according to schedule and planned performance. I find real happiness in this compared to previous missions. For MOM, we chose to play our genius," he has been quoted as saying in the magazine.

July 14, 2012, was a pleasant day in the historical city of Mysore in the state of Karnataka which was hosting the 39th meeting of Cospar, the Committee for Space Research, an international space organisation.

A large number of world space scientists had gathered at the Infosys campus which was the venue of the nearly one week meet.

Prior to the start of the conference, a media interaction was organised which was attended by newsmen from India, and some from abroad as well.

Following a brief introductory remark by Isro chairman, Dr Radhakrishnan, there came the usual volley of questions which he patiently answered.

During this meeting he made an important announcement which made headlines the next day: the much-awaited green signal from the government for India to launch an Indian mission to Mars was just around the corner.

His words came true because almost a fortnight later on August 3, 2012, the government gave its thumbs up.

Since then, he has been in command and ensured the fruition of this highly complex and challenging project in a span off just 15months, literally working 24X7.

A few days prior to the lift off, he and a few other members of the Mars team went and offered prayers at the famous Lord Venkateshwara temple in Tirupati and dedicated a model of the spacecraft to the Almighty.

It has been a long-standing tradition of ISRO scientists to visit this powerful temple and pray for the success of the mission.

In a recent post-launch interview he acknowledged that the Mars Orbiter Mission did confront him with some critical and nail-biting moments. Three of the most tough moments for him were the launch on November

5, 2013, and the trans-Martian insertion on December 1, 2013. The third difficult moment, he said, will be the Mars Orbit Insertion.

He explained that there were other challenges too such as the fine tuning of the last phase of the launch preparations which included harmonising the team and ensuring that the ship borne terminals--``SCI Nalanda'' and ``SCI Yamuna''-- were deployed in the South Pacific at the right time and position.

He said that the challenge was much more in this area considering that it was for the first time that ISRO was using civilian ships.

``We were anxiously awaiting a signal from `SCI Nalanda,' that all was progressing smoothly. So from the launch till the injection of the satellite into its initial earth parking orbit 42 minutes later was a challenging moment and also it was the longest flight by a rocket in ISRO,'' he explained.

``The in-orbit operations of the spacecraft were also a critical phase and we were suddenly faced with a technical problem,'' while pointing out that it was overcome successfully.

Regarding the possible exchange of MOM and Maven mission datas between the scientists of India and the US, he said that the joint space working group of both the countries could possibly work out something.

When not monitoring the flight to Mars, which is pretty rare, he listens to Carnatic music. This form of music belongs to Southern India. He is also a Kathakali dancer—a dance form of Kerala.

About future plans he said: "Yes, we have to do a major Mars mission, but the details have to be worked out."

MOM programme director, Mylswamy Annadurai, recalled in an interview about the launch on the afternoon of November 5. He said he had positioned himself in the spacecraft control centre at Bangalore. "In general, once the spacecraft is mated with the launcher all actions related to the satellite, shift to the satellite control centre. In particular after liftoff close monitoring of incoming signals from the satellite, and commands to the satellite all happen at this centre" he said.

Annadurai said as programme director and mission operations control board chairman, he had the responsibility along with mission director Dr. KesevaRaju and other team members at the spacecraft control centre to take care of the operations.

Recalled Annadurai: "For the spacecraft the first operation after separation from the launcher was the deployment of solar panels. It was a very crucial operation. We had to make sure that it happened, if not carry out the contingency plans. We had worked out detailed contingency procedures for the same. Sitting in front of the consoles waiting for the auto deployment was very anxious moment. That too the signal was coming from the ship borne terminal and this added some more anxiety," he said.

With regards to the Mars orbit insertion he said that enough simulations had been carried out. "Operation wise it is similar to all earth burns. Still we do plan to carry out one more mock up trial prior to the final Mars orbit insertion so that the operation takes place without any hitch. So confidence plus full preparedness is the answer," he explained.

He said that till the first week of March the spacecraft had consumed 535 kgs of fuel out of the total of 852 kgs.

Explaining how the Mars orbit insertion will be done on September 24, 2014 he said: "What we plan is to fire the 440N engine until we get the required delta-velocity. Nominally it may take around 30 minutes. All the commands required to prepare this firing in terms of the spacecraft orientation, when and where the firing should start, how long it should fire etc will be uploaded to the spacecraft well in time as was done during earth burns. The required orientations will be attained automatically as per these commands and firing will take place at the appropriate time and will continue until the required incremental velocity is achieved as monitored by the accelerometers onboard MOM."

" It is to be noted we will not be able to get this indication in real time due to two reasons, First, communication delay will be of the order of 20 minutes Second, a few seconds after the start of the engine's firing MOM will go behind Mars and will not be in the line of sight from the earth station." According to him the insertion will be the most significant and nail biting event after the launch.

He said that the families of the Mars team feel naturally a part of such missions as they have to make their own sacrifices to ensure their success.

Annadurai was the project director for Chandrayaan-1 and scientifically it was a thumping success---a fact acknowledged by NASA and other international space agencies.

Chapter 7

How the Mars Orbiter Mission was born

On May 11 1999, the auditorium of New Delhi's Ashok Hotel was filled with eminent personalities, politicians and media persons who had gathered to celebrate the first anniversary of Pokhran-2, the name given to the successful underground nuclear tests carried out by India exactly a year earlier.

The august gathering that evening was waiting to watch a presentation to be given by a person who had played a major role in India's space programme: Dr Krishnaswamy Kasturirangan, then chairman of ISRO.

Kasturirangan began by unveiling a chart which depicted various aspects of the nation's space development and explained its evolution---from the early sounding rockets to the present day Indian launch vehicles.

Dr. K. Kasturirangan
Member Planning Commission

FATHER OF THE INDIAN MOON MISSION
KRISHNASWAMY KASTURIRANGAN: CREDIT ISRO

He spoke about the Indian communication and remote sensing satellites and how they were benefitting the Indian public. In the course of his presentation, which nailed the audience to their seats, he raised a question which proved to be a major turning point in the history of the country's space missions: ``Has not the time now come for India to launch a planetary exploration?'' After this he threw a bombshell leaving a deep impact on the gathering. He declared that India was planning a mission to the moon!

His announcement left the audience speechless as it was the last thing they expected to be on the nation's space agenda. It took everyone by surprise setting off loud cheers and applause. What became evident was that at that moment the lunar mission had won instant support from the audience which perhaps reflected the response of the country too in the days ahead. It hit the headlines the next day triggering a mood of thrill and excitement almost in every part of the country, especially among the younger generation.

In the coming months the moon mission got the approval from various scientific bodies, Members of Parliament and more importantly Mr Atal Behari Vajpayee, who was then Prime Minister of India. He announced it during his Independence Day address from the Red Fort in New Delhi on August 15, 2003. He said with a lot of excitement: ``I am pleased to announce that India will send her own spacecraft to the moon by 2008. It is being named Chandrayaan.''

From the time Vajpayee made the announcement it took five years of preparations to launch the lunar flight. Then at sharp 6.22 a.m. on October 22, 2008, India's first unmanned scientific mission to the moon, Chandrayaan-1, lifted off at the Satish Dhawan Space Centre, Sriharikota. The rocket for this historic mission was the advanced version of the PSLV—the PSLV-XL. This was the inaugural flight of this version of the vehicle and it was the same one which carried the Mars Orbiter Mission.

TO THE MOON: INDIA'S FIRST LUNAR MISSION,
CHANDRAYAAN-1 LIFTING OFF AT SRIHARIKOTA IN THE
EARLY HOURS OF OCTOBER 22,2008: CREDIT ISRO.

There were 11 scientific instruments on Chandrayaan-1 of which five were from India abroad and six from abroad—two from NASA, three from the European Space Agency and one from Bulgaria. Among these perhaps the most important was India's Moon Impact Probe which crash landed with the Indian tri colour near the Shackleton Crater near the Lunar South Pole on November 14, 2008. The date was a sheer coincidence because the nation was celebrating Children's Day as it happened to be the birthday of former Prime Minister Jawaharlal Nehru who was very fond of kids.

The role of this probe is important because it discovered water on the moon. However, what has upset the Indian space community is that they have been not given credit for the important discovery.

CHANDRAYAAN-1 MOON IMPACT PROBE WHICH DISCOVERED WATER WHILE CRASH LANDING NEAR THE MOON' S SHACKLETON CRATER CLOSE TO THE LUNAR SOUTH POLE ON NOVEMBER 14,2008, BEING PREPARED FOR THE MOON MISSION.CREDIT ISRO.

Though the Indian moon mission had chalked up a number of accomplishments scientifically, it had to be unfortunately cut short by nearly a year following a technical problem. It lost contact with the earth on August 29, 2009 and the mission was abandoned.

All the same the lunar mission proved that India was technically capable of launching more challenging interplanetary missions. Despite the technical set back, it still boosted the morale and confidence of Indian space scientists and they became restless. The reason? Now, they wanted India to embark on a more challenging interplanetary mission. Therefore, after the scientific success of Chandrayaan-1, the question, therefore, was-- what should be the country's next destination?

Various options were evaluated and a consensus was finally arrived at---after the moon, India's next destination would be the Red Planet.

A year after Chandrayaan-1 concluded its mission; an initial move to head for the Red Planet was taken in August 2010 when ISRO set up a high powered team of scientists headed by Dr V. Adimurthy to study the feasibility to launch such a flight. It was called the Indian Mars Mission Study Team.

This group comprising eminent scientists stated that many space faring countries were launching Mars missions. It recommended that if India has to have a strong say globally in scientific, technological and strategic circles, there could be no other effective way than to launch a mission to Mars.

The team's report said that considering the orbital parameters of Earth and Mars there would be three chances to launch the flight—in November 2013, January 2016 and the third one in 2018.

The technical studies done by the scientists supported such a mission and the Space Commission gave the green signal. This commission oversees India's space programme and its approval is a must before a new mission is launched by ISRO.

The study team consisted of experts from different ISRO centres like Space Applications Centre, ISRO Satellite Centre, Vikram Sarabhai Space Centre, Satish Dhawan Space Centre, Liquid Propulsion Systems Centre, ISRO Telemetry, Tracking and Command Network, Indian Institute of Space Science and Technology, Physical Research Laboratory, Space Physics Laboratory and Laboratory for Electro-Optics Systems.

In an earlier interview given to the author for ``Mars Beckons India,'' Dr Adimurthy had stated that scientists were particularly curious to know about Mars because ``it holds the secrets of our past and the possibilities of our future.''

According to him, Martian science was expanding at a fast rate and many of the important questions relating to earth science translate to Mars science. 'A mission to Mars would provide an excellent opportunity to the scientific community to further understand the intricacies of Martian science, planetary formulation and evolution,'' he stated.

He explained: "Keeping in view the global scenario, gaps in Martian science and the existing capabilities to execute such a mission were all critically studied before proposing and formulating a mission to Mars."

He stated that the study team discussed topics like perspectives in Mars science, launch scenarios, detailed mission options for future launch opportunities, spacecraft design and the capabilities of the deep space network and finally concluded that India has the capability to aim for the Red Planet.

India's decision to go to Mars is not surprising because a natural follow-up to a moon mission in most space faring countries has generally been to aim for Mars. The reasons are both scientific and geo political. In the case of the latter a successful mission to Mars has always been a passport for a country to become a high profile member of the exclusive league of those countries having space programmes. These include the US, Russia, the European Space Agency, Japan and China.

That is not all. Geo politically a successful Mars mission will make the country's voice be heard in global forums like for example the UN and other important international meetings.

In this writer's earlier book about the India's Mars mission, "Mars Beckons India," he has quoted the well known film maker, Mr James Cameron who in his forward to Andrew Chaikin's book, "A Passion For

Mars,' has stated: "Mars has called us over the ages......Mars is a fantastic place, and the more we learn about it the more fantastic it becomes." Chaikin is a well known US-based space writer.

In his recent book "Mission To Mars," written by none other than the second man who stood on the surface of the moon, Buzz Aldrin, he says: "As always, front and centre is the power of Mars to entice us to brood over some key, compelling questions, particularly if life ever was sparked into being there. If so did it perish or is still resident on the planet?"

According to Aldrin "understanding the Martian climate and atmosphere, including the evolution of Mars's surface and interior can be looped back into grasping the past, present and future of Earth."

Aldrin, who has been championing the cause for a human mission to Mars points out that "continual scientific study of Mars is an important prelude to enable targeted, cost-effective human exploration. There's need to extensively characterise the surface and subsurface of Mars. Also, the polar regions of Mars are not only scientifically compelling, they merit study as resource-rich human destinations."

Robert Zubrin, the founder of The Mars Society, has described Mars as the Rosetta stone "for determining the prevalence and diversity of life in the universe." He has been a strong advocate for launching a manned mission to the Red Planet.

The main reason why nations want to invest on Mars missions is to find out if life ever existed on that planet since it is most earth-like. In fact this is also a goal of the Indian mission.

Another factor which is encouraging countries to go to the Red Planet is to understand the reason for its transformation from an early warm and wet state into the present cold and dry one.

A PICTURE OF MARS. THE ROLE OF MOST MARS MISSIONS IS TO FIND OUT IF THERE WAS LIFE ON THE PLANET, STUDY ITS ATMOSPHERE, LOOK FOR WATER AND FIND OUT WHY IT HAS TRANSFORMED FROM A WARM AND WET PLANET INTO A COLD AND DRY ONE: CREDIT NASA.

The other aim is to develop a detailed model of the planet incorporating the main data of its core, crust, the atmosphere keeping in view the main physical and chemical processes which are taking place on Mars.

An understanding of the Martian climate which has been changing, its atmosphere and the dust storms are among the reasons prompting exploration of Mars. Compared to earth's atmosphere, the one in Mars is rarified consisting of about 95 per cent carbon dioxide, three per cent nitrogen, 1.6 per cent argon and there are traces of oxygen and water.

According to this author's book ``Mars Beckons India,'' the Martian seasons are also of interest to the scientists. An analysis of its minerals containing mainly silicon, oxygen and metals is also being cited as a reason to launch Mars missions.

The ultimate aim of all these studies during an unmanned Mars mission is to explore the possibility of launching a manned flight and NASA has announced in its 2015 budget that it plans to implement this around 2030.

This is certainly not to say that India is planning a human flight to the Red Planet at this stage. Far from it, and right now at least such a massive project is nowhere in its horizon. But, in the years ahead if the country's technical capability further matures and strengthens, can such a mission be ruled out?

No.

Chapter 8

The mission costs a pittance for an individual!

A dinner at a good restaurant would cost more than Rs 1000. Or if you order a pizza at home it would mean spending about Rs 400 or perhaps even more. But, any guess how much MOM costs each individual on an average? Hard to believe, but just Rs four and that too it is a one-time payment!

On June 30, 2014, India's new Prime Minister, Narendra Modi, while speaking at Sriharikota after the successful launch of the PSLV carrying five foreign satellites, expressed happiness that the cost of the Indian Mars Orbiter Mission was much less than what it cost to produce the film ``Gravity,'' which was 100 million dollars.

According to a report in the US National Public Radio (NPR) ``India's Mars mission with a budget of 73 million dollars is far cheaper than comparable missions including NASA's 671 million dollars Maven spacecraft......'' which is among several publications noting the disparity between the cost of US space missions and India's burgeoning programme.

The total cost of the Indian Mars mission is Rs 450 crores.

The NPR report quotes MOM project director, S. Arunan, during a pre-launch media interaction as saying: ``this is less than one-tenth of what the US has spent on their missions. The cost-effectiveness of the mission is indeed turning out to be the highlight of the project, almost eclipsing other aspects," Arunan stated during the interaction.

The report states that his comments may have been directed partly at the critics of India's space programme who ``wonder why the country is spending 73 million dollars on interplanetary travel while millions of its people remain poor and malnourished."

Director of the Houston-based Rice Space Institute, David Alexander, has also been quoted by NPR saying there are some good reasons why India can do it cheaper than the US. ``It basically boils down to parts and labour," he said.

He remarked: ``I think labour is the biggest factor as well as the complexity of the mission. It takes a whole team of engineers," he explained.

Space engineers cost much less in India than they do in the US. Similarly, there's a vast gulf between the pay for electronic engineers. ``The average electronics engineer in the US makes a little more than 1,20,000 dollars as opposed to India, where he or she pull in less than 12,000 dollars annually at a company," Alexander stated in his interview to NPR.

NPR further says in general it seems safe to say engineers in India make between one-tenth and one-fifth of what their US counterparts do in absolute terms.

Cheaper And Quicker

Compared to its illustrious American counterpart, the Indian space agency delivers its missions in about one-third the time and at one-tenth the cost

Latest Moon Mission

ISRO		NASA
Chandrayaan	MISSION	Lunar Reconn-aissance Orbiter
2008	LAUNCH DATE	2009
18 months	TIME TO BUILD	3 years
$59 million	COST	$583 million

Latest Mars Mission

Mangalyaan	MISSION	MAVEN
Nov 5, 2013	LAUNCH DATE	Nov 18, 2013
18 months	TIME TO BUILD	5 years
$69 million	COST	$671 million

Source: ISRO, NASA

Alexander has called India's mission to Mars as a very complex one, more difficult than Chandrayaan-1 five years ago. "Mars missions have a history of failure," he stated, while pointing out that fewer than half of them launched by the US, Russia and the European Space Agency have been successful.

He said no country has reached Mars on the first try. China's 2011 attempt with Russia to send the Yinghuo-1 probe fizzled out when the Russian rocket failed to leave earth orbit. A 2003 mission by Japan got farther, but couldn't get into Mars orbit.

Reuters points out a successful Indian mission will have the effect of "positioning the emerging Asian giant as a budget player in the latest global space race."

Describing the mission as a triumph in low cost engineering, well known Indian aerospace scientist, Roddam Narasimha, told the The New York Times (NYT) in a recent interview that "by excelling in getting so much out of so little, we are establishing ourselves as the most cost-effective centre globe wide for a variety of advanced technologies."

NYT says that even a priority sector like space research in India gets a meagre 0.34 per cent of the country's total annual outlay. "Its one billion dollar space budget is only 5.5 per cent of NASA's budget," the paper has stated.

It states that India's abundant supply of young technical talent helped reign in personnel costs to less than 15 per cent of the budget.

The modest budget of the mission, the paper said, did not allow for multiple iterations. "So instead of building many models like a qualification model, a flight model and a flight spare, as is the norm for American and European agencies, scientists built the final flight model right from the start. Expensive ground tests were also limited," NYT states.

According to the paper, ISRO reduced the cost of the mission by transforming old technology into new which in this case happened to be the PSLV rocket.

The cost was kept to the minimum by using similar systems in other space projects. "Systems like the attitude control, which maintains the orientation of the spacecraft, the gyro, a sensor that measures the satellite's deviation from its set path, or the star tracker that orients the satellite to distant objects in the celestial sphere are the same across several Isro missions," it has stated.

Highlighting the significance of the Indian Mars mission, Krishnaswamy Kasturirangan said that planetary exploration forms an important component of country's space ambitions.

Pointing out that this has to be executed step-by-step, he told this author recently that ``it not only demonstrates our capabilities, but in more ambitious programmes it will strengthen our credibility while other countries consider international collaboration with India.''

Surprisingly, the mission has been praised by even a supposedly arch rival of India like China. For example, Ye Hailin, an expert on South Asian studies at the Chinese Academy of Social Science said that the Indian mission to Mars should be rationally interpreted as `` a great achievement of India that also deserves applause from the rest of the world.''

``Like the Chinese, Indian people have their space dreams as well. The Mars orbiter, if successful, will increase the human race's store of knowledge and change our life,'' he said

The success of the launch on November 5 prompted Beijing to call for a joint effort by both the countries to ensure peace in outer space. Chinese foreign ministry spokesperson, Hong Lei, said that outer space is shared by the entire mankind. ``Every country has the right to make peaceful exploration and use of outer space,'' he said.

The remarks of the Chinese assume significance in the context of their maiden mission to the Red Planet being unsuccessful. The Mars-bound 115 kg spacecraft, Yinghuo-1 with four scientific instruments was launched on November 8, 2011, by a Russian Zenit rocket along with the Russian Phobos-Grunt sample return spacecraft.

The role of Yinghou-1 was to study mainly Mars's surface and atmosphere. But the trans-Martian insertion failed as a result of which both Yinghuo-1 and Phobos-Grunt got stranded in the orbit. Thereafter, both spacecraft underwent a destructive re entry on January 15, 2012 and finally disintegrated over the Pacific Ocean.

India has time and again emphatically clarified that the country is not in a race with China to reach Mars. This was the repeated response from the Indian government following the decision of Isro to fast track the Mars mission in just 15 months. This fast tracking led many foreign space experts and political analysts to wrongly speculate that India was only trying to beat China in the race to the Red Planet.

The mission has left a strong impact on the people of India, especially among students. It has inspired the younger generation to take a deeper interest in science, technology, engineering and mathematics.

MOM mission director P. Kunhikrishnan during the launch said ``capturing and igniting the young minds of India and across the globe will be the major return of the mission."

This writer had interacted with the students of the Bombay International School in December about the Mars mission and in March spoke at a conference organised by the international students' space body called the Students For the Exploration and Development Of Space (SEDS) at the Sastra University in Thanjavur, Tamilnadu on the same topic and the response was one of enthusiasm.

In many schools and colleges the mission has become a favourite topic for a classroom project.

What better proof of this than a film about the mission being produced for a change not by professionals, but by amateurs. And who were they? A group of 15 enthusiastic students of the prestigious K.R. Mangalam World School at Gurgoan in the state of Haryana not far from New Delhi.

The film made by youngsters from classes V to 1X was screened at the National Science Film Festival in February at Bangalore. Of the 105 entries, it was selected among the 30 best.

Called ``Mission Mars,'' the 14-minute movie is in the form of a talk show which attempts to dispel misconceptions about the amount spent on the mission.

It involves various characters like news reporters, a team of scientists and common people. It revolves around the technological importance of the mission, possibility of life on Mars, human settlement on the Red Planet and the capital involved in the mission.

This film has attracted a lot of praise and is being shown at various forums. It is playing a key role in furthering the interest and awareness of the flight in the country.

Chapter 9

What Next?

Will India stop with just a single orbiting mission to the Red Planet?

It is unlikely.

Indications are that a second Mars mission is already on the cards.

The current flight is a technology demonstrator which is mainly to establish that India has the capability to enter the orbit around Mars.

The second mission to Mars is expected to have more scientific content.

In fact in December 2013, none other than director of ISRO's Vikram Sarabhai Space Centre, Dr. S. Ramakrishnan, publicly declared that the space agency ``may go in for a more sophisticated sequel to the Mars mission by 2016 if the Geo Synchronous Satellite Launch Vehicle (GSLV) is ready.''

The centre which is located at Thiruvanathanapuram is the main complex for the design and development of launch vehicles.

Ramakrishnan, however, clarified that the second one has yet to be approved." But we are planning a second mission to Mars with a more powerful launch vehicle in two years. The GSLV will be ready by then and also the GSLV Mk-111 version," he said.

He even went a step further to say that the proposed second flight unlike the first will have a lander. " The final approval to the mega project will depend upon the performance of MOM," he has been quoted as saying.

The GSLV did have did have a successful flight after four failures with an indigenous cryogenic engine on January 5,2014, raising hopes that it could be the rocket for the second Indian mission to the Red Planet.

But, just one successful flight of the GSLV with an indigenous cryogenic engine is not enough to convince the space fraternity that it was ready to carry a spacecraft with a lander to Mars. So, there has to be more trial flights before the GSLV can be declared operational and relied upon for a challenging mission, like for example flying a heavy Mars-bound spacecraft having more scientific content.

Let us not forget that till January 5, 2014, the GSLV was called the ``naughty boy" because of its erratic record. On January 5 it became a ``good boy" and it has to retain this title.

The Indian space fraternity will be disappointed if the country's Mars mission programme concludes with just one flight. They are very keen and enthusiastic for a second one and perhaps even more.

That is not all. India's second mission to the moon, Chandrayaan-2, which is expected to be launched between 2016 and 2018, has a lander and a rover developed indigenously.

It seems unlikely that ISRO will develop just one lander and rover and then halt the programme. More landers and rovers will be there on the drawing boards of the scientists and it is believed that some of them could be heading for Mars in future.

According to NASA, each Mars mission is part of a continuing chain of innovation. Each relies on past missions for proven technologies and contributes its own innovation to future missions.

It says that so far the exploration of Mars occurred in three stages: fly-bys, orbiters, landers and rovers. Future NASA missions envisage the use of aircraft, balloons, subsurface explorers and sample return missions. Cannot India emulate this example?

The European Space Agency's Mars Express which was launched on June 2, 2003 has paved the way for the next generation of European-led Mars exploration mission, namely ExoMars.

After the Chinese inability to get to Mars in its first attempt in November 2011 on account of the failure of the Russian rocket, China has confidently declared that it ready to launch another mission to the Red Planet.

When this author interviewed Dr Radhakrishnan for his book ``Mars Beckons India," the latter hinted at the possibility of a second Indian mission to Mars having more scientific content.

In the course of the interview he also assured that the entire country and perhaps also the whole world will join India on its first mission.

His words proved true because on October 22, 2013, ISRO launched an official Mars Orbiter Mission face book which proved instantly popular. Congratulatory messages poured in from different countries when the mission successfully lifted off on November 5, 2013.

Therefore, judging from the response in the Face book it is amply clear that if India does not launch more missions to Mars, not only will the people of India feel let down, but those abroad too.

So what is the final conclusion? India by no means should stop with just one orbiting mission to Mars. There should be landing ones too with rovers. Such missions perhaps will go a long way if at all India is planning to team up with other nations in embarking on a complex Mars sample return mission.

India has to think big with a larger vision, if at all it has a dream of one day embarking on a human flight to the Red Planet!

Just imagine the day when the people of India hear that the Indian tri colour has landed on the surface of Mars as it happened on the moon on November 14, 2008

That will be the day!

Chapter 10

From the Church to the Red Planet

It was in every way an auspicious launch for India's space programme.

The lucky take off was neither from a laboratory nor a workshop, but from a beautiful church located in the fishing village of Thumba, not too far from Thiruvanathanapuram airport.

The situation had a touch of irony because a hi-tech sector like space exploration had its early beginnings in a traditional environment of a church dedicated to St Mary Magdalene. And where else could such a beginning take place except perhaps in India?

It all started in the late 50s and early 60s when two eminent Indian scientists, Dr Vikram Sarabhai who took the early steps to launch India into the space era, and Dr Homi Bhabha, the father of the country's nuclear programme, surveyed different locations in southern India for setting up a rocket launching station.

Among the places they visited was Thumba. After evaluating various sites, they finally zeroed down on Thumba, which at that time consisted of, apart from the church, a beach, coconut trees and a few houses. That is all. The silence of the place was punctured by the sound of the waves lashing the shore.

What swung the decision in its favour was the fact that it was near the earth's magnetic equator. Dr Sarabhai and Dr Bhabha were convinced that Thumba's proximity to the earth's magnetic equator would help considerably for doing atmospheric research.

But, it was not exactly a trouble-free acquisition of the place. Taking over the nearly 600 acres for making a rocket launching station would mean displacing the local population and relocating them nearby to their convenience.

Understandbly, therefore, when the fishermen first heard of the plan there were murmur of mild protests, one of the main reasons being that they would be away from the church.

Dr Sarabhai and Dr Bhabha spent anxious moments apprehending that the villagers may not give up the place. If this happened it would be a replay of an earlier scenario----both of them would have to carry out another survey to identify an alternative location which would considerably delay the launch of the country's stellar journey.

But, the then Bishop of Thiruvanathanapuram, the Right Reverend Dr Peter Bernard Periera, asked them to set aside their fears assuring that there was every chance of the issue being settled amicably.

The Bishop, who backed the space programme, arrived in Thumba on a Sunday morning and addressed the fisherman at the church. Emphasising the important role space technology played in the country's development, he said it would individually benefit them too in the days ahead.

With these words he asked the congregation how many of them would support India entering the space era? His statement which finally clinched the deal was that the nation's path to the stars would begin in their own village and it would thus become internationally famous.

Hearing him they got excited. The villagers discussed among themselves briefly and then came their much-awaited verdict. And, how did answer? They said in a chorus unitedly, Amen! What did it mean? Yes, they have all agreed to support the plan of Thumba being handed over to the space department for making it a rocket launching station.

Thus the Bishop by opening the doors of the Church to the Indian space department had also opened the doors of the space sector to this country.

With the blessings of the fishermen India's entry into the space era had now been assured marking it an auspicious start.

Once they gave their approval, the process of taking over the area was finished in about 100 days. After it was completed scientists, including Dr APJ Abdul Kalam, who later became the President of India, moved in to Thumba to set up the facilities.

The Genesis

St. Mary Magdelene Church in the fishing village of Thumba in Thiruvananthapuram, Kerala served as the main office for scientists in the early stages of India's space programme. It's preserved in its full glory even today and it houses an impressive space museum.

This church is certainly the Mecca of Rocket science in India.

This was in the early 60s and rocket components and payloads were carried by bicycles and bullock carts. The facilities were minimum and were in stark contrast to the scenario today.

The scientists, totally dedicated to the challenging task of making India a major global space player never complained about working long hours under harsh conditions. For instance the task of satisfying their hunger at the end of a gruelling day was a major problem, and to make matters worse they stayed in a poorly-furnished hotel at Thiruvanathanapuram.

When they returned to Thiruvanathanapuram after a hard day's work at Thumba, which was usually late at night, most of the cafeterias in the city were closed leaving the scientists no choice but to grab whatever was available at the snack bar of the Thiruvanathanapuram railway station.

Their relentless efforts began to pay dividends, and Thumba slowly expanded with new facilities including a launch pad and even a mission control centre. The centre was named as the Thumba Equatorial Rocket Launching Station or TERLS from where sounding rockets would be launched.

Sounding rockets are used for carrying out atmospheric studies and Dr Kalam and some of his team members were trained at Nasa's Wallop Island in the US for designing, developing and launching these type of rockets.

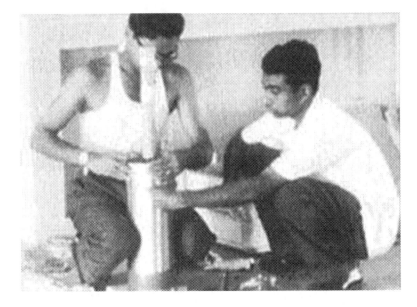

THE BEGINNING: DR KALAM AND HIS COLLEAGUE
WORKING ON A ROCKET AT THUMBA: CREDIT ISRO

When various types of construction activity had almost been completed, preparations began for launching the first sounding rocket from Thumba.

But since the design of an Indian sounding rocket had yet to get off the ground, it was apparent that the first one to be launched will not be an indigenous vehicle, but an imported one.

Accordingly, a sounding rocket called Nike-Apache was acquired from Nasa which was launched on November 21, 1963 at 6.25 p.m. Its launch marked India's entry into the space era giving enough confidence to the scientists to design and develop indigenous sounding rockets.

November 21,2013, nearly a fortnight after the launch of MOM, marked the 50th anniversary of the launch of the first sounding rocket at Thumba. The first India-made sounding rocket, the RH-75, lifted off from the station in November 1967.

In the last 50 years this centre has witnessed the take off of more than 2000 sounding rockets, many from the US, the UK, Japan, France and Russia.

On May 11,2012, Asia's first student-designed and developed sounding rocket called "Vyom," was successfully launched. It was built by the students of Isro's Indian Institute of Space Science and Technology also at Thiruvanathanapuram.

This rocket launching station, once a sleepy fishing village, has now turned into a beehive of activity dealing in critical areas like rocket design, propellants, motor casting, integration, payload assembly, evaluation and building sub systems.

The church has been made into a space museum attracting hundreds of visitors all through the year.

As the sounding rockets were notching up one success after another, Dr Sarabhai became convinced that if India had to become a major player in the global space sector, then it has to develop rockets having the capability to place satellites in orbit. In short he felt that the time had come for this country to acquire a satellite launch capability.

He spoke to the scientists about his vision and all of them enthusiastically backed him. With this was born the satellite launch vehicle programme and the first rocket—a four stage one-was designated as the SLV-3. Of the four design options, the third one was chosen and so it was called SLV-3.

The D-day was August 10,1979 and the attention of most parts of the world was on India. Why? The first Indian rocket with a satellite launch capability was being readied for lift off at Sriharikota with a 40 kg Rohini spacecraft.

The launch time was fixed for 7.58 a.m. All was a 'go.' This flight would open a new chapter in the history of India's space programme.

The countdown was progressing smoothly, but when it hit the eight second mark ie when there were just eight seconds left for the launch, the computer indicated that a technical problem had been detected in the rocket and the mission was a 'no go,' atleast for the moment.

Confronted with the grim warning, the scientists went into a snap session to evaluate the problem. After a detailed discussion they reached the consensus---the problem was a minor one making them confident that it would not jeopordise the mission. Dr Kalam who was the mission director endorsed the decision and gave the 'go' for launch.

At sharp 7.58 a.m. the rocket took off setting off a huge round of applause by the scientists in the mission control room.

It rose higher and higher and the first stage was functioning flawlessly. The scientists clapped and cheered watching the rocket and had no doubts it would be a successful flight.

But, their sense of confidence gradually gave away to one of uncertainty and once again they became tense when the second stage of the rocket ignited. At this point, the performance of the vehicle became erratic and it started veering off its designated flight path.

The rocket went out of control and exactly 317 seconds after launch the mission had to be terminated. Instead of zooming towards the sky, the country's first rocket with a satellite launching capability splashed into the Bay of Bengal nearly 560 kms off the coast off Sriharikota with the satellite.

A post flight analysis stated that the failure was the result of a flaw in the rocket's second stage control control system.

This major setback cast a pall of gloom in Sriharikota causing many to become demoralised.

Despite their frustration and disappointment they at the same time stood by Dr Kalam and assured him of their full support. The top guns of the Indian space fraternity asked Dr Kalam not to consider the failure a setback and even went to the extent of asking him to start preparing for the next flight.

July 18,1980 dawned over Sriharikota. There was hectic activity which was marked with a sense of nervous apprehension among the scientists. This was the day which had been fixed for the next SLV-3 flight carrying the Rohini satellite.

As the sun rose over the Bay of Bengal the scientists sat glued to their computers in the mission control room praying that this time the flight should be a success. The question in their minds was whether the rocket will head for the sky or the sea?

At 8.03 a.m., the rocket thundered off the launch pad gathering velocity every second. Then 12 minutes after launch at 8.15 a.m. Dr Kalam who was the mission director made the most important announcement of his life. He declared: "Mission director calling all stations. Stand by for an important announcement. The fourth stage apogee motor had given the required thrust to put Rohini satellite into orbit." The flight was a success and at last India had rocketed into the space era in the true sense. The prayers of the scientists were answered.

The success at once unleashed a sense of joy and excitement not only at Sriharikota, but all over the country as well. Dr Kalam became an instant national hero!

THE SECOND SLV-3 A FEW MINUTES BEFORE LAUNCH ON JULY 18,1980. ITS MISSION WAS A SUCCESS USHERING INDIA INTO THE SPACE ERA. CREDIT ISRO.

There were three more flights of the SLV-3, the last one on April 17,1983. Their success gave confidence to the scientists to go in for a more advanced version of the rocket which was the five-stage solid propellant Augmented Satellite Launch Vehicle (ASLV).

The project was initiated in the early 80s, the role of the rocket being to place 150 kg class of satellites in the low earth orbit. Unfortunately, this vehicle had a chequered record with, of the four launches between 1987 and 1994, only one being completely successful, one a partial success and there were two failures.

The ASLV was considered a bridge between the SLV-3 and another rocket which became the work horse of Isro and proved a super success. It is the four-stage 44.4 metre tall Polar Satellite Launch Vehicle (PSLV).

The PSLV can place 1600 kg class of satellites in the 620 km sun-synchronous orbit and those weighing 1050 kg in the geo stationary transfer orbit.

There are three variants of this rocket—the standard type, the core alone (without strap ons) and the XL which is an advanced version of the vehicle.

The advanced one was used for India's maiden lunar mission, Chandrayaan-1 and the Mars Orbiter Mission.

India wanted to reduce its dependence on foreign countries to launch its communication satellites. With this in mind it initiated project in 1990 to design a three-stage 49-metre tall rocket called the Geosynchronous Satellite Launch Vehicle (GSLV) for launching Insat(Indian National

Satellite System)-class of communication satellites weighing between 2000 and 2500 kg in the geostationary transfer orbit.

As stated earlier, the rocket had a spotted record because of the eight missions only three were successful and one was a partial success.

Of the failed flights one was on April 15,2010 which was the first mission to be powered by an indigenous cryogenic engine. Eight months later on Christmas Day a GSLV plunged into the Bay of Bengal because of a fault in the Russian system.

This rocket became a victim of geopolitics which resulted in its development suffering a setback.

The programme envisaged Russia supplying seven cryogenic engines to power the upper stage of the rocket. But, the US citing the Missile Technology Control Regime managed to temporarily stall the deal between India and Russia. The American apprehension, which was totally baseless, was that India would use the cryogenic engines to power its missiles!

But, after hectic negotiations between India and Russia, and Russia and the US, the deal was revived and the Russian cryogenic engines reached India. As they arrived Isro also began the development of the indigenous cryogenic engine to power the GSLV.

As stated earlier, this ``naughty boy" became a ``good boy," on January 5,2014, when a GSLV with an indigenous cryogenic engine operated successfully raising hopes that this rocket would be the one for future interplanetary missions.

In the days ahead if this rocket retains its reputation of being a ``good boy" with more trial flights it will be finally declared operational. This means India need not depend upon Arianespace to launch its two-tonne class of communication satellites from Kourou in French Guyana.

Infact if this vehicle proves dependable, even foreign agencies can consider using this rocket for launching their communication satellites.

A newer version of the GSLV designated as GSLV Mark 3 is expected to be test flown later this year, its primary role being to launch communication satellites weighing between 4500 and 5000 kg. Once declared operational, India need not approach Arianespace to launch satellites of this category.

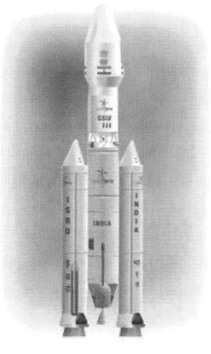

THE GSLV MARK 3. THE ROCKET WHICH MAY TAKE AN
INDIAN TO SPACE FROM SRIHARIKOTA: CREDIT ISRO.

And if India decides on a more ambitious exploration of the solar system in future maybe this will be the rocket which will be in the picture.

The GSLV Mark 3 is also being considered for the Indian human space flight programme, though this project has yet to receive the formal green signal from the government.

The programme involves two Indians flying in the low earth orbit for about a week and then landing in the sea. Significantly, the crew module for this mission made by Hindustan Aeronautics Limited in Bangalore was delivered to Isro recently. This module will be tested during the trial flight of the GSLV Mark 3.

Isro's 2025 vision document includes planetary exploration and also the solar system and launching a manned space programme among its key factors.

This document has now to be seen in the background of the statement made by Dr Sarabhai when he initiated the Indian space programme in the 60s saying: ``We do not have the fantasy of competing with the economically advanced nations in the exploration of the moon or the planets or manned space flight.''

Has not his vision and the programme deviated from the original trajectory chalked out by Dr Sarabhai in the 60s?

Yes, it certainly has, and undoubtedly for the right reasons too.

Had it not moved away from Sarabhai's original vision, the chances of India becoming a major global space power would have been slim.

Ends

INTERNATIONAL MARS MISSIONS

Launch Date	Name	Country	Result	Reason
1960	Korabl 4	USSR (flyby)	Failure	Didn't reach Earth orbit
1960	Korabl 5	USSR (flyby)	Failure	Didn't reach Earth orbit
1962	Korabl 11	USSR (flyby)	Failure	Earth orbit only; spacecraft broke apart
1962	Mars 1	USSR (flyby)	Failure	Radio Failed
1962	Korabl 13	USSR (flyby)	Failure	Earth orbit only; spacecraft broke apart
1964	Mariner 3	US (flyby)	Failure	Shroud failed to jettison
1964	Mariner 4	US (flyby)	Success	Returned 21 images
1964	Zond 2	USSR (flyby)	Failure	Radio failed
1969	Mars 1969A	USSR	Failure	Launch vehicle failure
1969	Mars 1969B	USSR	Failure	Launch vehicle failure
1969	Mariner 6	US (flyby)	Success	Returned 75 images
1969	Mariner 7	US (flyby)	Success	Returned 126 images
1971	Mariner 8	US	Failure	Launch failure
1971	Kosmos 419	USSR	Failure	Achieved Earth orbit only
1971	Mars 2 Orbiter/Lander	USSR	Failure	Orbiter arrived, but no useful data and Lander destroyed
1971	Mars 3 Orbiter/Lander	USSR	Success	Orbiter obtained approximately 8 months of data and lander landed safely, but only 20 seconds of data
1971	Mariner 9	US	Success	Returned 7,329 images
1973	Mars 4	USSR	Failure	Flew past Mars
1973	Mars 5	USSR	Success	Returned 60 images; only lasted 9 days
1973	Mars 6 Orbiter/Lander	USSR	Success/ Failure	Occultation experiment produced data and Lander failure on descent
1973	Mars 7 Lander	USSR	Failure	Missed planet; now in solar orbit.

1975	Viking 1 Orbiter/Lander	US	Success	Located landing site for Lander and first successful landing on Mars
1975	Viking 2 Orbiter/Lander	US	Success	Returned 16,000 images and extensive atmospheric data and soil experiments
1988	Phobos 1 Orbiter	USSR	Failure	Lost en route to Mars
1988	Phobos 2 Orbiter/ Lander	USSR	Failure	Lost near Phobos
1992	Mars Observer	US	Failure	Lost prior to Mars arrival
1996	Mars Global Surveyor	US	Success	More images than all Mars Missions
1996	Mars 96	Russia	Failure	Launch vehicle failure
1996	Mars Pathfinder	US	Success	Technology experiment lasting 5 times longer than warranty
1998	Nozomi	Japan	Failure	No orbit insertion; fuel problems
1998	Mars Climate Orbiter	US	Failure	Lost on arrival
1999	Mars Polar Lander	US	Failure	Lost on arrival
1999	Deep Space 2 Probes (2)	US	Failure	Lost on arrival (carried on Mars Polar Lander)
2001	Mars Odyssey	US	Success	High resolution images of Mars
2003	Mars Express Orbiter/ Beagle 2 Lander	ESA	Success/ Failure	Orbiter imaging Mars in detail and lander lost on arrival
2003	Mars Exploration Rover - Spirit	US	Success	Operating lifetime of more than 15 times original warranty
2003	Mars Exploration Rover - Opportunity	US	Success	Operating lifetime of more than 15 times original warranty
2005	Mars Reconnaissance Orbiter	US	Success	Returned more than 26 terabits of data (more than all other Mars missions combined)
2007	Phoenix Mars Lander	US	Success	Returned more than 25 gigabits of data

2011	Mars Science Laboratory	US	Success	Exploring Mars' habitability
2011	Phobos-Grunt/ Yinghuo-1	Russia/China	Failure	Stranded in Earth orbit
2013	Mangalyaan	India	En route	On way to Mars
2013	Mars Atmosphere and Volatile Evolution	US	En route	On way to Mars

Source: NASA'S JET PROPULSION LABORATORY